AF405388

The Technical Learning Series

BOOK 1

Understanding Industry

Core Concepts

ANSWER BOOKLET

Seth Dolcy

Copyright © 2022, 2023 Seth Dolcy

All rights reserved.

No parts of this book shall be copied, scanned, printed or transmitted in any way without the express, prior, written consent of the author(s).

ISBN: 978-976-96814-5-3

Caribbean Examinations Council [R], Caribbean Secondary Education Certificate [R], CXC [R] and CSEC [R] are registered trademarks of the **Caribbean Examinations Council** and are used by permission.

Editing / Layout by
Passionate Words Editing Services

(IG - @passionate.words.editing246)

Preface

This is the companion Answer Booklet to the
Understanding Industry: Core Concepts Manual. It is
divided into the same chapters and sections as the
Manual, so that the answers to the questions posed at
the end of each section can be easily located.

It is our hope that these resources will aid the student in
their understanding of the subject matter in
preparation for their Caribbean Secondary Education
Certificate ® (CSEC ®) examinations as produced by the
Caribbean Examinations Council ® (CXC) ®

TABLE OF CONTENTS

CHAPTER FOUR: ELECTRICAL INSTALLATION AND ELECTRONICS.

CHAPTER FIVE: TYPES OF WORK SITE WASTE.

CHAPTER SIX: Standards For Fire Prevention, Response And Types Of Hazards & Practicing Occupational Health, Safety And Welfare Standards.

CHAPTER SEVEN: EVACUATION PRACTICES

CHAPTER EIGHT: TYPES OF WORKSHOP HAZARDS.

CHAPTER NINE: SOCIAL AND ECONOMIC IMPACT OF INDUSTRIES.

CHAPTER ONE:
CONSTRUCTION AND MANUFACTURING INDUSTRIES.

1.1 - IDENTIFYING THE AREAS WITHIN THE CONSTRUCTION AND MANUFACTURING INDUSTRIES.

1.   Define the term "industry" in your own words.

An **industry** can be defined as the economic activity that focuses on the processing of raw materials, the manufacturing of goods and the availability of services that serves a need in society.

2.   State two (2) differences between the construction and manufacturing industries?

The **construction industry** refers to the design and construction of physical structures such as buildings or roadways, while the **manufacturing industry** would focus on the production of finished goods sold to distributors, retailers, or consumers. Therefore, the main difference would be that construction is focused on taking a building from initial concept to a completed physical structure and manufacturing is the mass production of goods and items.

3. What are the benefits and disadvantages of manufacturing?

The **manufacturing** industry's benefits are:

- Provides employment
- Technological improvements
- Reduces import expense
- Stabilises the economy
- Reduces the production time

The **manufacturing** industry's disadvantages are:

- Environmental issues
- Lacks creativity
- Potential to increase unemployment

4. Name two (2) types of structures construction is used to build

Two types of structures that can be built in using construction are: (examples)

- Office buildings
- Houses
- Supermarkets
- Hotels

1.2 - THE CONSTRUCTION SUBDIVISIONS

1.   **What are the main characteristics that distinguish a residential dwelling from a commercial area?**

The main characteristics that would distinguish a **residential dwelling** from a **commercial area** is that the residential dwelling is what a person would call their home whereas the commercial area is where products are bought and sold.

2.   **Give two (2) examples of what businesses would be found in an industrial park.**

Examples of businesses that would be found in an **industrial park** are as follows:

- Warehouses
- Distribution centres
- Food services
- Clothing
- Factories

3.   State the job description of a civil engineer?

The job description of a **civil engineer** is to strategize, design, and oversee massive projects.

4.   State the reasons why a civil engineer is important to a project.

A **civil engineer** is paramount to a project because they:

- Work alongside the architect to design structures.
- Ensure that each project that is being undertaken is environmentally safe.
- Address how to manage traffic effectively.
- Exceed the expectations of the public with the assistance of public officials.

1.3 - THE MANUFACTURING SECTORS

1.   State two (2) reasons why the manufacturing sector is important in your country.

The **manufacturing sector** is needed since:

- It has helped in modernising agriculture and the manufacturing of tool and equipment.
- Halting unemployment and poverty.
- Growth in the economy.

2.   Name a manufacturing sector that was not mentioned above and describe what it does.

The following are additional manufacturing sectors:

- **Computer and Electronics**- This sector produces computers, communications equipment, and similar electronic products. Also, they produce many electronics products or components that are found in other sector's products, such as cars, entertainment, and appliances.
- **Transportation**- This sector provides the necessary services required to transfer people or goods in addition

to the infrastructure needed to facilitate transportation.

- **Furniture**- This sector is involved with the creation, manufacturing, allocation, and sale of functional and decorative objects and household items.

3.   **State two (2) ways in which the manufacturing sector can stabilize an economy.**

The **manufacturing sector** can stabilize an economy by

- Creating employment
- Providing revenue
- Encouraging spending
- Reducing poverty

4.   **In your own words, state what is meant by manufacturing.**

Manufacturing is the means to which goods and items are produced by the processing of raw materials by tools, human labour, or machinery.

1.4 - ORGANIZATIONAL STRUCTURE OF THE CONSTRUCTION AND MANUFACTURING INDUSTRIES.

1.	Create an organizational chart of personnel (managers, workers, task, and relationships) that work in the construction and manufacturing sectors.

Refer to Figures 3-5 in the Manual.

2.	In what type of sector (construction or manufacturing) would the following structures be most suited for; 1) top-down structure, 2) flat structure, and why?

- A top-down structure describes a traditional organizational style that emphasizes the vision of upper management. This is best suited for both the construction and manufacturing sector.

- The flat organizational structure is when the employees have a higher level of responsibility in the organization. This organizational structure is commonly used by smaller businesses or those who would rather a more modern approach to management.

3.   **Give two (2) reasons why a suitable organizational structure must be selected for a sector.**

The reason why a suitable organizational structure should be selected for a sector is because:

- The organization needs to have a level of stability.
- The best management techniques should be utilized.
- It would give responsibility to those who would require it.

4.   **If you had a construction company, which structure would you select, and why?**

A matrix structure is one of the most applied organizational structures for a construction company. A matrix structure is a combination of functional and project organizational forms that can best guide the organization.

1.5 - OCCUPATIONAL LEVELS AND THEIR FUNCTIONS IN THE CONSTRUCTION AND MANUFACTURING INDUSTRIES..

1.   Why are skilled workers a necessity in the construction and manufacturing industries?

The reason why skilled workers are a necessity in the construction and manufacturing industries is due to the need for professionalism, a high level of workmanship and accountability.

2.   Why should professionalism be paramount in the construction and manufacturing industries?

Professionalism is paramount in the **construction** and **manufacturing** industries because professionalism focuses on having a reliable, dependable, and efficiently run company.

3.   There is a shortage of skilled workers in your country. How would you remedy this? Discuss.

To remedy the shortage of skilled workers, the country will have to establish a technical vocational school to certify the workers.

4.   List four (4) skills a worker should exhibit on a construction site when problem solving issues, and why?

A worker should exhibit the following skills when problem solving

- Critical thinking.

- Communication skills.

- Building inter-personal relationships.

- Influencing skills.

The reason why these skills are important for the worker is that it allows him/her to be able to address and adequately deal with the issue amicably.

1.6 - BUILDING AND MANUFACTURING CODES AND STANDARDS.

1.   State two (2) benefits of following codes related to the respected industries.

The following of **industry codes** would result in safety and quality.

2.   How would the use of standards in an industry affect it?

The use of **standards** can indeed be positive for an **industry** with reliability, consistency, compatibility, and compliance of the product, service or material is ensured.

3.   What is the difference between a code and a standard?

The differences between codes and standards are that a **code** instructs what needs to be done, and a **standard** states how to do it. A code may say that a building must have a lighting system. The standard will show what kind of system is needed and how it must work.

4.   **Give examples of two (2) codes and two (2) standards found in construction and manufacturing.**

Examples of codes and standards in construction and manufacturing are as follows:

Codes: Building Code and Caribbean Uniform Building Code.

Standards: ASTM International standards, ISO standards and National Electric Code (NEC) standards.

1. **How do qualifications assist in the career paths of construction and manufacturing workers?**

A qualification can assist in the career paths of construction and manufacturing workers since it highlights to the employer that the worker has the knowledge and understanding required to successfully complete the tasks handed to them. In addition, qualifications provide professional development for the worker which can potentially lead to higher positions.

2. **How are tradespersons overlooked in construction and manufacturing industries?**

The role of the **tradesperson** can never be overlooked since they have an important part to play in the success of the project. Tradespersons are the workers that complete the physical work and should be respected as such.

3.	An Architect is considered the person that designs, develops, and supervises projects. How does the Architect's role differ from that of the upholsterer?

Placing the label of "greater importance" on workers can have a damaging affect. The role of the architect is important and so is the upholsterer's. So to say the that one is more important than the other would depend of the need of their skills and expertise.

4.	Do you believe that the CVQs are a productive move for the Caribbean currently? Explain why.

The CVQs is a productive move for the Caribbean at this time due to the need for CARICOM workers to be trained from a practical standpoint rather than an academic one.

CHAPTER TWO:
Entrepreneurship

2.1 - FUNDAMENTALS OF ENTREPRENEURSHIP

1.   What is the difference between a small business and an entrepreneur?

The difference between a **small business** and **entrepreneurship** is that a small business is one that is local and does not have a large amount of funds whereas an entrepreneur can start as a small business but that is not the total vision.

2.   Define "goal setting".

Goal setting is when a person focuses on the things that are important to their goals by not concentrating on the things that would hinder them.

3.   Why is realism important to an entrepreneur?

Realism is important to an entrepreneur because within the concept of goal setting, decisions about what objectives should/ must be achieved (for example, wealth or growth) should be realistic and achievable.

4.   Are entrepreneurs needed at this time? Explain why.

I do believe that entrepreneurs are needed since the world is consistently changing and there is always a need for new thinking and foresight in business to keep up with those changes. Entrepreneurs create new opportunities with the businesses they create.

2.2 - A BUSINESS PLAN CONCEPT FOR AN ENTREPRENEUR

1. **Why is it that government legislation is important to the entrepreneur? Discuss.**

Government legislation assists the entrepreneur by ensuring that they are covered by the laws and the guidelines of the nation.

2. **Is there a safer alternative to the lending institution? And if so, what would you consider to be a better option?**

A safer alternative to the lending institutions would be to form a group of investors to buy into the business.

3. **Do the risks associated with with a lending institution outweigh the benefits? Discuss.**

I do not believe that the risks associated with a lending institution outweigh the benefits, but I do believe that without careful planning to repay debt it can cause a massive issue in the long run.

4.   **Would it be wiser for an entrepreneur to contact a bank or credit union for a loan? Why?**

It would be smarter to do the necessary research to ascertain which lending institution can offer the better benefits between the two options.

2.3 - GOVERNMENT LEGISLATION FOR ENTREPRENEURSHIP

1.   Design a business (name, description, purpose, etc.) from the construction or manufacturing industries.

Name: Dolcy Industries

Description: Dolcy Industries are the makers of finely crafted furniture for all households.

Purpose: To provide the customers with the highest quality workmanship for their hard-earned money.

2.   What are the important factors for creating a business plan for your company?

The creation of a business plan allows the business to be able to

- Track their course towards success
- Set goals
- Assess objectives

3.   **Are business plans helpful? If so, how?**

A business plan sets the objectives, goals, expenditures, and purpose of the business. This serves as the means to assess the success of the business.

4.   **Create a business plan for a business (construction or manufacturing) of your choosing to compete in your local market.**

1.   Name of Company: DOLCY INDUSTRIES

2.   Contact Number: 1-246-437-8000

3.   Name of Applicant: MR. S. DOLCY

4.   Is this a new or existing business?

☒   New   ❏ Existing Date business established:

____

5.   Type of Operation:

❏ Manufacturing   ❏ Wholesale/Retail

❏ Agriculture   ☒   Construction

❏ Aquaculture / Fishing   ❏ Transportation

❏ Forestry   ❏ Mining

❏ Tourism   ❏ Other (Specify)

----

6.   Mailing Address: TREE ROAD, ST. MICHAEL

7.   Postal Code: 12345

8.   Telephone Numbers:

Residence: 1-246-432-6754   Cell: 1-246-245-8756

Fax: 1-246-435-6790   E-mail:

CONSTRUCTIONPLUS@GMAIL.COM

9. Proposed location of business: TREE ROAD, ST. MICHAEL

10. Legal Form of Business:

❏ Incorporated   ☒   Sole Proprietorship

❏ Partnership   ❏ Co-operative

❏ To be incorporated

11.   Date Incorporated: ________________________________

2.4 - IDENTIFYING ENTREPRENEURSHIP OPPORTUNITIES IN THE CONSTRUCTION AND MANUFACTURING INDUSTRIES

1.   **What are the advantages and disadvantages of entrepreneurship?**

Advantages of entrepreneurship

- Being your own boss
- Defining your income
- Generating income

Disadvantages of entrepreneurship

- Not guaranteed a paycheck
- No off days
- Stress levels can be high

2.   **Critically assess which upcoming construction or manufacturing business would be best suited at this time to be introduce into the current market?**

The construction and manufacturing business that would be best suited during this pandemic would be

- Home made products

- Industrial cleaning
- Home improvement kits

3.   There is a difference between a small business and entrepreneurship. Discuss.

Yes, I do believe that there is a difference between a small business and entrepreneurship. A small business focuses on a known and established product and/or service, while entrepreneurship deals with a new, innovative product or service.

4.   What would be the best course of action for an individual if they are interested in exploring entrepreneurship?

If entrepreneurship is a path to be undertaken, it would be best to gain an understanding in

- Identifying the right business for you.
- Determine if you should be educated for your business
- Plan your business approach
- Find the group/audience you want target
- Networking
- Sell your idea to investors
- Know your market

2.5 - ENTREPRENEURIAL OPPORTUNITIES THAT LEAD TO SELF-EMPLOYMENT, PRODUCTIVITY AND WEALTH CREATION.

1.   How can an entrepreneur contribute to the economy?

An entrepreneur can contribute to the economy with the creation of jobs and their innovations.

2.   What would motivate an entrepreneur?

What motivates an entrepreneur is the fact that they are providing products and/or services that are needed by the nation that in turn affords a source of income and sustainability.

3.   Is entrepreneurship sustainable in a pandemic?

I do believe that entrepreneurship is sustainable in a pandemic since there should be innovative technologies that should be undertaken to better cope with the changes that have occurred in the world. The entrepreneur has to be creative and seek to establish him/herself by focusing on what the needs or wants in the market are at the time.

4. What are the entrepreneur's greatest obstacles?

The entrepreneur's greatest obstacles are:

- Inadequate training for what needs to be undertaken
- The need for human resources to assist with the business.
- The fear of not being successful
- The lack of the necessary practical knowledge..

2.6 - ENTREPRENEURIAL OPPORTUNITIES THAT LEAD TO SELF-EMPLOYMENT.

1.   What are your thoughts on entrepreneurship opportunities during a pandemic? Would you start a business during this time, if so, why, or why not?

The current pandemic has the potential to halt the progression of any entrepreneurship opportunities. However, with the correct amount of planning with something like a business continuity plan, it would work. There is nothing wrong with starting a business during this time but it would be best to have an idea what can be profitable and what the public needs especially with social distance as a factor.

2.   What business would you start given the opportunity at this time in the construction or manufacturing sectors?

I would start a business in making wooden alcohol dispensers made from mahogany. In the current climate, sanitization is imperative; the dispensers would be exactly what businesses and individuals could use to effectively sanitize the environment for both customers and staff.

3. Having done research, which innovation in the construction or manufacturing industry have you found to be the interesting and progressive?

The innovation that I found to be the most interesting and progressive in the construction industry is self-healing concrete.

4. Do you agree with the statement, "Technology has become the backbone of both the construction and manufacturing industries?" Discuss.

I agree with this statement since technology has been essential to the continuation of both the construction and manufacturing industries with the various innovations that can been created.

I disagree with the statement because technology has not been able to make the impact in the respective areas other than in the use of computers.

CHAPTER THREE
STANDARDS FOR FIRE PREVENTION, RESPONSE AND TYPES OF HAZARDS & PRACTICING OCCUPATIONAL HEALTH, SAFETY AND WELFARE STANDARDS.

3.1 - ELECTRICAL FIRE HAZARDS

1.   **Critically address the impact of a fire accident on a work site.**

Accidental fires or explosions can cause devastation in terms of death, injuries, damage to property and the environment.

2.   **Fire extinguishers are extremely important to the efficiency and safety of the workshop. What are the other ways in which you can extinguish a fire?**

The basic techniques used to extinguish a fire are:

- To smother the fire by limiting its access to oxygen.
- To cool it with water which reduces the heat.
- Remove the fuel or oxygen source.

3.   **Which extinguisher would you use to extinguish a Class A and Class C fire?**

Dry powder extinguishers are suitable for fighting burning solids, liquids and gases which are considered Class A and C fires.

4.   What are the proper steps to follow to operate a fire extinguisher?

The easy-to-remember four (4)-step process to operating a fire extinguisher can be achieved by the PASS method:

- **PULL** the pin located at the top of the fire extinguisher, while holding the extinguisher away from you to unlock the mechanism.
- **AIM** low toward the base of the fire to extinguish it from the source.
- **SQUEEZE** the lever slowly to avoid any mishaps.
- **SWEEP** the nozzle from side to side at the base of the fire to spread over a wider area.

3.5 - HEALTH AND WELLNESS STANDARDS

OBSERVING PERSONAL HYGIENE AND APPEARANCE STANDARDS

1.   Why is personal hygiene important to an individual?

Personal hygiene is important to an individual because with a high standard of hygiene it is possible to avoid the development and spread of infections, illnesses, and bad odours.

2.   What does self-esteem have to do with body image?

Self-esteem has a lot to do with body image since a person who views their body in a negative or positive way can affect their self-esteem in the same manner. A good body image is needed to better function around others in a productive way.

3.   Why is personal hygiene important to a construction worker?

Personal hygiene provides the necessary confidence needed to be comfortable at work and around fellow workers.

4.   How does proper hygiene promote good health?

With the consistent practice of proper hygiene, an individual is better able to combat the risk of infection and possibly death.

3.6 - PRACTICING WELLNESS/FITNESS PROGRAMME

1. **What effective measures would you put into place to address conflict on the work site?**

The effective measures that can be put in place to better address conflict as it arises in the workplace could be finding out what the issues are; knowing time and place to bring up disagreements and listening to the opinions of others.

2. **In your opinion, how can stress affect a worker and his productivity?**

Stress can cause a worker to not be able to concentrate and fully apply themselves to the work at hand. This can cause the lack of productivity and disruptions in the workplace.

3. **How would you show empathy to others in your workplace?**

Empathy can be shown to those in your workplace by being

able to understand the feelings and emotions of those around you.

4. How effective is conflict management in the workplace?

Conflict management is extremely important since this method of management is used to ensure that the business is able to function with issues that can disrupt productivity.

3.7 - MANAGING INTERPERSONAL CONFLICTS.

3.8 - HOW DO YOU MANAGE INTERPERSONAL CONFLICTS?

1.   What are the possible benefits of having interpersonal relationships with your co-workers?

Having interpersonal relationships with your co-workers allows for a level of companionship to develop that can lead to a greater level of teamwork and productivity.

2.   Conflicts are most common under what circumstances in the workplace?

Conflicts can be common in certain circumstances such as the behavioural component, cognitive component, and the effective component.

3.   State two (2) of the reasons why workers disagree when they are engaged in activities with each other?

These are various reasons why co-workers would have some level of disagreement

- Bad communication

- Unable to follow instructions

- Failure to respect one another

- Not having the same goals and objectives

4.   How would you handle a disagreement while working as part of a team? Provide an example.

There was a project to complete with some co-workers. The path to be taken to complete the project was not clearly explained but the goal was established. This caused the team not to be able to work together due to the uncertainty of the proper approach to successfully finish the project. However, after further conversations on the matter, a resolution was found, and the previous disagreements were dealt with.

3.9 - ENVIRONMENTAL SAFETY PRACTICES

1.   **What would you consider to be a safe working environment?**

A **safe** working environment is one that allows the worker to be comfortable and able to work without the fear of being injured.

2.   **Explain what environmental safety practice is.**

Environmental safety practices are the policies, and procedures that ensure the safety and well-being of those persons in the immediate area while proper waste disposal, containment and storage of toxic chemicals is being done.

3.   **Why should work related policies be made available to employees?**

Work related **policies** should be made available to employees because the employees have the right to know what procedures are to be undertaken in the case of workplace hazards.

4.   **List three (3) environmental safety issues.**

- **Electrical** hazards are those that happen in the handling of electricity.

- **Physical** hazards occur naturally in the environment.

- **Chemical** hazards can be both considered natural and human-made chemicals in the environment.

3.10 - SAFETY AND MAINTENANCE STANDARDS

1.   **What is workshop safety?**

Workshop safety is defined as the guidelines required for the those who work in the workshop to be able to work safely and comfortably.

2.   **What are four (4) possible requirements of the workshop?**

The four requirements of the workshop are:

- To only have portable tools that normally operate at 110 volts.
- Have clearly identified pedestrian routes.
- Have pits that are covered when not in use or have guard rails in place.
- Have an inspection scheme for all tools and equipment so that they are safe to use.

3.   **Why is a checklist important for safety in the**

workshop?

A checklist is important for safety in the workshop because it allows the workers to know the rules and regulations to be adhered to in the workshop.

4.   How can a workshop be an area of risk?

The workshop can be an area of risk due the hazards that can happen as a result of carelessness and failure to follow the proper use of tools and equipment.

3.11 - HEALTH AND WELLNESS STANDARDS

1.   What affect do health issues have on the workplace?

Health issues have the potential to disturb how the workplace functions.

2.   What are safety programs?

Safety programs seek to provide the necessary guidelines that would highlight the correct procedures (including checklists) which allow the work environments to be safer by preventing accidents. In addition, workers must know the safety procedures that need to be followed in their specific area.

3.   Describe a wellness program.

Wellness programs mainly include certain activities such as weight loss programs, exercise, stress management programs, and wellness assessments that are designed to help workers to eat better, lose weight and improve their physical health.

4.   How do health and wellness standards assist the workplace?

Health and wellness standards assist in the workplace by giving the workers the means to address the workers' physical, mental, and work-related issues. Therefore, the worker is better able to be productive in the workplace.

3.12 - BASIC FIRST AID STANDARDS

1.   **What is first aid?**

First aid is when emergency or immediate care is rendered to a person who is injured or has fallen ill until professional medical treatment is made available.

2.   **Who should administer first aid to someone who is injured?**

The best person who should administer first aid is a person who is trained in first aid.

3.   **Where should a first aid kit be located?**

A first aid kit should be in an area where it is easily accessible and in clear sight.

4.   **List four (4) items found in the first aid kit and their functions.**

- **Disposable sterile gloves**- to be used when the patient is bleeding

- Tweezers- to remove large to medium sized obstructions found in a cut.

- Scissors- used for cutting through various thin materials

- Alcohol-free cleansing wipes- used for cleaning cuts, grazes, and light-wounds.

3.13 - TREATING OF MINOR BURNS, ELECTRIC SHOCKS, WOUNDS AND BLEEDING, ABRASIONS, INJURIES TO BONE (STRAINS, SPRAINS)

1.   **Burns are classified by degrees from first to fourth. Describe a third degree burn.**

A **third degree burn** destroys two full layers of your skin and instead of the skin turning red, it may appear in various colors such as black, brown, white or yellow. This type of burn will not cause pain due to damaged nerve endings.

2.   **Electrical burns can be caused by current found in the home or workshop, certain batteries, and even lightning. What should be done first after a person has an electrical burn?**

The first thing to be done after a person has received an electrical burn is to be sure that the person is not in contact with the electrical source anymore. In addition, locate the source of the current and turn it off. Also, remove the current from the person by using a material that is a nonconductor of electricity, such as wood, plastic, or cardboard, etc.

3.   What would be the first procedure to follow if someone is bleeding in the workshop?

The first procedure to follow would be to apply firm pressure on the cut or wound with a clean cloth, tissue, or piece of gauze until the bleeding has stopped.

4.   Describe what is the difference between a strain and sprain?

The difference between a **strain** and a **sprain** is that while a **strain** involves an injury to a muscle or to the band of tissue that attaches a muscle to a bone, a **sprain** tends to refer to the injury of the bands of tissue that connect two bones together.

3.14 - PRACTICING RECOVERY POSITION AND MOUTH-TO-MOUTH RESUSCITATION

.1.   You are walking on the street on your way home. There is an elderly lady in front of you and she collapsed. You ran to her side to see if you can assist in anyway. You checked her vitals, and they were fainted. You realized at this moment that she was not breathing. What would you do?

- Ensure the lady is lying on a hard, flat surface.
- Look into her mouth and throat to ensure that the airway is clear.
- Tilt the head back slightly to open the airway.
- Place your mouth tightly over her nose and mouth. Blow two quick, shallow breaths. Watch for her chest to rise.
- Next remove your mouth and look for the chest to fall as she exhales.
- Listen for the sounds of breathing. Feel for the lady's breath on your cheek. If breathing does not start on its own, repeat the procedure.

3.15 - PROCEDURES FOR REPORTING AN ACCIDENT AND GETTING ASSISTANCE.

1. **Please find below a completed injury report for the workshop.**

I am reporting a work related: ☒ Injury ☐ Illness ☐

Near miss

Your Name: John Smith

Job title: Student

Teacher: Mr. S.Dolcy

Have you told your teacher about this injury/near miss? ☒

Yes ☐ No

Date of injury: 13 / 10 / 2021

Names of witnesses (if any): W. Black and T. Thompson

Where, exactly, did it happen? In the workshop.

What were you doing at the time? I was using the tool incorrectly I was using the chisel and mallet to cut pieces for my project and was talking to a fellow student and not paying attention.

Describe step by step what led up to the injury.

I was talking to Tom and wasn't paying attention and the chisel slipped and cut my finger.

What could have been done to prevent this injury?

I should have been paying attention to my work.

What parts of your body were injured? My left hand index finger

Did you see a doctor about this injury/illness?

☐   Yes   ☒   No

If yes, whom did you see?   Doctor's phone number:

Date: Time:

Has this part of your body been injured before?

☐   Yes   ☒   No

If yes, when? Supervisor:

Your signature: J. Smith   Date: 13/10/2021

3.16 - BASIC EMERGENCY RESPONSE STANDARDS

1.   **Why is a risk assessment necessary when creating an emergency response plan?**

A **risk assessment** is necessary in the creation of an emergency response plan because an understanding of what can happen will make it possible to determine correct resources required to develop plans and procedures for preparation.

2.   **What should be placed in an emergency response plan for a workshop?**

The **emergency response plan** should include:

- The possible emergencies, consequences, required actions, written procedures, and the resources available so the workers would be adequately prepared.

- A detailed lists of emergency response personnel which includes cell phone numbers, alternate contact details, and their duties and responsibilities in case of emergency.

- Updated floor plans of area.

- Map/s showing evacuation routes.

3.   **How to deal with a situation when someone is injured in the workshop?**

If a person is injured in the workshop, here are a few suggestions to deal with the situation:

- Take control of the area the accident occurred.
- If possible, administer first aid and have someone call for emergency services for help if possible.
- Deny access to people who don't need to be in the area.

4.   **Why is it important to know the emergency numbers?**

The importance in knowing the emergency numbers is that these numbers are solely used to save lives and by remembering them it is possible to save your life and others.

3.17 - PRACTICING SAFETY STANDARDS FOR WORKSHOP AND WORK SITE.

1.	List ten safety rules for both the workshop and work site.

1. Always wear safety equipment
2. Wear appropriate clothing
3. Keep work shop clean
4. Check equipment before use
5. No drinking and eating in the workshop
6. No playing in the workshop
7. Stay away from all hazards in the workshop
8. When in the workshop make sure there is always supervision
9. Any faulty of equipment report to supervisor
10. Never use blunt blades and bits

2.	**Explain why workshop rules are put in place.**

The reason why workshop rules are established is because without the necessary guidelines to follow there will be an increase in the injuries that would occur in the workshop. The workshop rules are to ensure that the workers know the procedures that will allow them not to be harmed.

3.   Why should a workshop be cleaned after work is done?

The purpose for cleaning the workshop after work has come to an end is to ensure that there are no hazards, the work area is clean, and it is ready for when work commences again.

4.   Who is responsible for workshop safety?

Workshops are under the control of a supervisor but more importantly it is the responsibility of everyone to ensure that the workers can operate in a manner that encourages the use of safe work practices.

CHAPTER FOUR:
ELECTRICAL INSTALLATION AND ELECTRONICS.

4.1 - LOCAL AND REGIONAL STANDARDS.

1.   What is the result when electricians do not follow the standards and regulations of electrical installation?

By failing to follow the standards and regulations of electrical installation, the electrician is liable to cause severe damage to homes or properties or indirect/direct injury to individuals.

2.   Why is qualification and certification important for the electrician?

Qualification and certification are important for the electrician because being qualified means the individual went through the proper schooling. In addition, when a person is certified that shows that they have earned a specific skill and gained a particular set of knowledge.

3.   **Why regional standards differ from country to country?**

The reason why regional standards would vary from country to country is due to different infrastructures, environments and demands of the individual country.

4.   **What polices do the IEEE and NEC standards implement?**

The **IEEE standards** are dedicated to advancing innovation and technological excellence for everyone and the **NEC standards** are there to ensure that electrician installations are reliable, approved, and up to code.

1.   How can an electrical shock occur?

An electrical shock can happen when an individual comes into contact with both wires of an electrical circuit.

2.   Why should the wires that make up the circuit be of the correct type?

The reason why the wires of the circuit should be of the correct type is because if the incorrect wires are used it can cause a fire.

3.   List (3) three practices when using electrical equipment that can potentially reduce the risk of injury or fire.

The three practices that reduce the risk of injury or fire while using electrical equipment are:

- Use of electrical equipment in cold rooms must be minimized due to condensation issues.
- Disconnecting the power source before servicing or repairing electrical equipment.
- If contact is met with a live electric wire, do not touch

the equipment, source, cord or individuals.

4. Name two (2) of the daily hazards a telecom worker would face.

Two of the daily hazards a telecom worker would encounter would be falling from great heights and electrocution.

CHAPTER FIVE:
TYPES OF WORK SITE WASTE.

5.1 - ASSESSING AND CONTROLLING HAZARDOUS SUBSTANCES-SPILLAGES AND LEAKAGES OF CHEMICALS AND OTHER HAZARDOUS SUBSTANCES.

1.   Name (2) two complex spills that can occur on a work site.

The complex spills that can occur on a work site are:

- Lead
- Paint
- Pesticides
- Petroleum products

2.   Why should workers be trained in the spillage cleanup procedures?

The reason why workers should be trained in the spillage cleanup procedures is because they would have the knowledge and able to make the correct judgment call for the situation at hand.

3.   List the type of PPE should be used during a spill in the workshop?

The following are the PPE that should be used during a spill in the workshop:

Safety helmets, gloves, eye protection, hazmat suits, high-visibility clothing and safety footwear.

4.   How would a simple spill be cleaned up in the workshop?

First, assess the spill to find out what type of chemicals are exposed. Next, the most suitable PPE is selected. Follow the procedure of the workshop in cleaning up the spill. Lastly, properly dispose of the spilled chemicals.

5.2 - WASTE DISPOSAL METHODS.

1.   **Why is waste disposal important to the well being of the construction site?**

Waste disposal is important to the construction site because waste can become a hazard to the workers and consume valuable space.

2.   **List (3) three types of construction site waste.**

The types of construction site waste are:

- Concrete,
- Bricks,
- Tiles,
- Cement, and
- Ceramics.

3.   **What are the benefits of the correct waste disposal methods being used?**

The benefits of proper waste disposal can include a high standard of health and safety for the workers and having a good reputation in the community.

4.   How can construction site waste be created?

Construction site waste can be created when demolition is done or a spill occurred and the chemical cannot be reused.

5.3 -RECYCLING METHODS.

1.   List the PPE equipment that a worker can wear to be protected when handling tertiary recycling materials?

The PPE equipment that a worker can use when handling tertiary recycling materials would be the following:

Disposable face masks, visors, disposable gloves, ear plugs, hair nets, safety glasses and disposable garments etc.

2.   Which materials used on a construction site can be recycled?

The materials that are used on a construction site that can be recycled after use can be the following:

Wood (treated and untreated), concrete, rebars, glass etc.

3.   Why use recycling methods on construction sites?

The reason why to use recycling methods on construction sites is that recycling has the potential to save money, improve the surroundings of the work site and prevent hazards.

4.   **What factors would decide if a worker used a recycling or waste disposal methods?**

The factors that would decide if a worker would use a recycling or waste disposal methods is how hazardous the materials are and if it would be practical to reuse them.

CHAPTER SIX:
STANDARDS FOR FIRE PREVENTION, RESPONSE AND TYPES OF HAZARDS & PRACTICING OCCUPATIONAL HEALTH, SAFETY AND WELFARE STANDARDS.

6.1 - ENGINEERING PRODUCTION.

1.   **Why should codes and standards be adhered to?**

The reason why codes and standards should be adhered to is because the codes and standards are in place to ensure that people and their property are not damaged because of not following the recommended safety procedures.

2.   **With the introduction of the various institutions and their methods of operating regarding codes and standards, how can this be beneficial to the industry?**

The industry needs the intervention of the various institutions for the governing of the codes and standards because it enables the industry to function collectively with uniformity and professionalism.

3.   In your opinion, why are the institutions that deal with the codes and standards are non-profit organizations?

In my opinion, the reason why the institutions are non-profit organizations is because non-profit organizations are created for services. This designation does not permit the organization to be a source of income or profit, thus allowing the authentic and unbiased information to be produced.

4.   How can the local government contribution to the formation of an industry standard?

The local government can contribute to the formation of an industry standard by having input to what is required for the nation and its betterment.

CHAPTER SEVEN:
EVACUATION PRACTICES

7.1 - RESPONDING TO EVACUATION ALARM SOUNDS.

1.	**What is an evacuation plan?**

An escape plan is a plan for a group of people that indicates how to exit a building and identifies an assembly spot in the event of an emergency.

2.	**Why is it important to determine an assembly area outside?**

It is important to have an **assembly area** since it can ensure that everyone is accounted for, to prevent anyone from returning inside and to indicate to firefighters that no one is inside.

3.	**Who should create an evacuation plan?**

The emergency response officer/s is/are the person/s that

should draw up an evacuation plan.

4.   What should a person take when evacuating?

At this time, it would be best to be concerned with yourself and your personal possessions.

7.2 -PERFORMING EVACUATION ROUTE MAPS, GATHERING POINTS AND BULLETIN BOARDS.

1.   Should emergency evacuation maps be a requirement in a workshop?

I do believe that evacuation map should be a requirement in a workshop because the workers need to know where they are and how to get to certain locations.

2.   What are exit routes?

An **exit route** is a continuous and unhindered path of exit from any point within a building to somewhere that is safe.

3.   Where should the emergency evacuation maps be posted?

The emergency evacuation should be placed those areas that workers have clear sight of to remind them of the emergency evacuation plan.

4.   What is the purpose of an emergency evacuation plan?

An emergency evacuation plan seeks to provide the workers with information that enables them to evacuate a building as quickly and as safely as possible.

7.3 - PERFORMING EMERGENCY PROCEDURES FOR FIRES AND NATURAL DISASTERS (hurricanes, earthquakes, floods, tsunami, volcanoes)

1.   **What is the purpose of the Incident Management Team (IMT)?**

The purpose of the IMT is to ensure that trained persons are available to respond to emergencies

2.   **What is considered the freelance period?**

The freelance period is when the responders can act independently of supervision.

3.   **What is personnel accountability?**

Personnel accountability seeks to improve the safety of emergency workers by keeping track of their locations and assignments when operating at an incident site.

4.   **Who should do authorizing during larger incidents?**

The person who should be designated during a large incident is the **accountability officer**.

CHAPTER EIGHT:
TYPES OF WORKSHOP HAZARDS.

8.1 - TOP FIFTEEN (15) WORKSHOP HAZARDS.

1.   **Which word best describes the likelihood of an incident occurring?**

The word that best describes the likelihood of an accident occurring is **hazard**.

2.   **What is the first step to take when conducting a risk assessment?**

The first step to be taken when conducting a risk assessment is identifying the hazards.

3.   **Why is it important that all workers in their daily job role have the appropriate training?**

The reason why it is important is that it would assist in decreasing the hazards in the workplace.

4.   **Why is it good practice to report near misses regarding accidents?**

It would be considered good practice to report near misses because this can stop a more serious event occurring in the future.

8.2 - HOW TO REDUCE HAZARDS IN YOUR WORKSHOP.

1.   Address the affects unidentified hazards can have on a worker and business.

Unidentified hazards can potentially cause harm to a worker in the future and place the business under undo financial strain.

2.   Define risk assessment.

Risk assessment is described as the overall process or method where hazards are identified that have the potential to cause harm to the workers.

3.   Safety is everyone's business; how can a worker effectively protect themselves from machinery and hazards?

A worker can effectively protect themselves from machinery and hazards by being properly trained and aware of his/her surroundings.

4.   What are the responsibilities of the worker?

A worker must take reasonable care of themselves and not do anything that would affect the health and safety of others at the workplace.

CHAPTER NINE:
SOCIAL AND ECONOMIC IMPACT OF INDUSTRIES.

9.1 - TRADE BLOCKS.

1.   What are the advantages of CARICOM for Caribbean nations?

 The advantages of CARICOM are through functional cooperation in education, health, culture, and security.

2.   What does the 12 member states of CSME have in common?

The 12 member states of the CSME are from the Caribbean.

3.   Why does the member states need funding from international organizations?

The member states need the funding from international organizations because the Caribbean countries require the influx of funds to better able to achieve their goals.

4.   In your opinion, why isn't Cuba a signatory in the agreement with the other nations?

Cuba isn't a signatory in the agreement with the other nations because Cuba does not want to be part of the funding process.

www.ingramcontent.com/pod-product-compliance
Lightning Source LLC
Chambersburg PA
CBHW071228130726
47998CB00002B/866